U.S. Department
of Transportation
**National Highway
Traffic Safety
Administration**

www.nhtsa.gov

DOT HS 811 375

October 2010

The Effectiveness of Underride Guards for Heavy Trailers

DISCLAIMER

1. Report No. DOT HS 811 375	2. Government Accession No.	3. Recipient's Catalog No.
4. Title and Subtitle The Effectiveness of Underride Guards for Heavy Trailers		5. Report Date
		6. Performing Organization Code
7. Author(s) Kirk Allen, Ph.D.		8. Performing Organization Report No.
9. Performing Organization Name and Address Evaluation Division; National Center for Statistics and Analysis National Highway Traffic Safety Administration Washington, DC 20590		10. Work Unit No. (TRAIS)
		11. Contract or Grant No.
12. Sponsoring Agency Name and Address National Highway Traffic Safety Administration 1200 New Jersey Avenue SE. Washington, DC 20590		13. Type of Report and Period Covered NHTSA Technical Report
		14. Sponsoring Agency Code

15. Supplementary Notes

16. Abstract

Federal Motor Vehicle Safety Standards (FMVSS) Nos. 223 and 224 require underride guards meeting a strength test on trailers with a GVWR of 10,000 pounds or greater manufactured on or after January 24, 1998. FMVSS No. 224 defines the size requirements for the guards, while FMVSS No. 223 describes strength testing and energy absorption requirements for DOT-compliant guards.

This report is a statistical analysis of crash data aimed at determining the effectiveness of FMVSS-compliant underride guards at preventing fatalities and serious injuries in crashes where a passenger vehicle impacts the rear of a tractor-trailer. The primary findings are the following:

- Data from Florida and North Carolina showed decreases in fatalities and serious injuries to passenger vehicle occupants when rear-ending a tractor-trailer subsequent to the implementation of FMVSS 223 and 224. However, the observed decreases are not statistically significant at the 0.05 level, possibly due to the small sample sizes of the data.
- Using supplemental data collection from North Carolina, it is shown that passenger vehicle passenger compartment intrusion is more apt to occur when the corner of the trailer is impacted, rather than the center of the trailer. This result is statistically significant at the 0.01 level.
- It is not possible to establish a nationwide downward trend in fatalities when a passenger vehicle rear-ends a tractor-trailer – neither in terms of total number of fatalities, percentage of fatalities in rear impacts relative to other passenger vehicle fatalities involved in tractor-trailer accidents, nor number of fatal crashes per 1,000 total crashes. The Fatality Analysis Reporting System does not list the model year of the trailer.

17. Key Words NHTSA; NCSA; FARS; State Data System; underride; heavy vehicles; tractors; trailers	18. Distribution Statement Document is available to the public from the National Technical Information Service www.ntis.gov		
19. Security Classif. (Of this report) Unclassified	20. Security Classif. (Of this page) Unclassified	21. No. of Pages 41	22. Price

Form DOT F 1700.7 (8-72)   Reproduction of completed page authorized

Executive Summary

This report is a statistical analysis of several crash databases to determine the effectiveness of underride guards at preventing fatalities and injuries in crashes where a passenger vehicle impacts the rear of a tractor-trailer.

NHTSA mandates that all trailers with GVWR of 10,000 pounds or greater manufactured on or after January 26, 1998, be equipped with an underride guard. The dimensional requirements are specified in FMVSS No. 224 (Rear Impact Protection), while the strength testing and energy absorption requirements are outlined in FMVSS No. 223 (Rear Impact Guard).

The Truck Trailer Manufacturers Association (TTMA) issued a voluntary Recommended Practice RP 92-94, *Rear Impact Guard and Protection* in April 1994, including the dimensional requirements of the subsequent NHTSA standard (FMVSS 224) but lacking the energy absorption requirement (FMVSS 223). From 1952, to 1998, trailers and semi-trailers were Federally regulated by Federal Motor Carrier Safety Regulations that mandated rear impact guards, but allowed substantially smaller guards than the NHTSA standard and the TTMA recommended practice. These standards also allowed the guards to be set further from the rear of the trailer and imposed no strength tests on the guards.

Several data sources were analyzed to determine the effectiveness of FMVSS-compliant underride guards, relative to the guards in use prior to model year 1994, in terms of preventing fatalities and serious injuries. Using the Fatality Analysis Reporting System (FARS), it was not possible to determine a reduction in passenger vehicle fatalities. The model year of the trailer is not recorded in FARS, thus no precise analysis can be performed in terms of the type of underride guard on the trailer. The total number of passenger vehicle fatalities when rear-ending a tractor-trailer has not decreased over the years in terms of total number of fatalities, percentage of fatalities in rear-end impacts relative to other passenger vehicle fatalities involved in tractor-trailer accidents, or in number of fatal crashes per 1,000 total crashes.

Two data sources contain the model year of the trailer – data from Florida in NHTSA's State Data System and data from North Carolina in a special data collection project coordinated through the State Highway Patrol. There are reductions of passenger vehicle fatalities and serious injuries in rear-end collisions with trailers subsequent to the implementation of FMVSS 223 and 224, relative to pre-standard trailers, and relative to other collisions between passenger vehicles and tractor-trailers. The results from Florida are based on more crashes from a greater number of calendar years than the results from North Carolina. In Florida, there is an observed 27 percent reduction in fatalities for trailers that should be equipped with FMVSS-compliant underride guards and a 7 percent reduction in fatalities or serious injuries; however, neither reduction is statistically significant at the 0.05 level. External factors were considered but did not appreciably change the effectiveness calculation – these included roadway classification, passenger car versus LTV, restraint use in the passenger vehicle, and age of the tractor portion

ii

(power unit) of the tractor-trailer. The results from North Carolina are numerically higher than the results from Florida, but also not statistically significant at the 0.05 level – an observed 83 percent reduction in fatalities (marginally significant at the 0.10 level) and a 57 percent reduction in fatalities or serious injuries. These results are more uncertain because there is much less data than in Florida, thus extraneous factors are not accounted for.

Taken together, the results from Florida and North Carolina are consistent with expectations at the time of publication of the Final Rule that FMVSS-compliant guards would be responsible for a reduction in fatalities and, to a lesser extent, serious injuries to passenger vehicle occupants. But they are not unequivocal evidence that the guards are effective and the FARS analysis shows that there remains room for improvement.

In North Carolina, supplementary information was collected about the angle of impact and the extent of passenger vehicle compartment intrusion. It is clear that impacts into the rear corners of the trailer are more prone to passenger compartment intrusion compared to impacts into the center rear portion of the trailer, regardless of the type of underride guard. For FMVSS-compliant guards, the proportion of crashes with intrusion when a passenger vehicle impacts the rear center of a trailer is significantly lower than for crashes in which a passenger vehicle impacts the rear corner of a trailer. The result is statistically significant at a 0.01 level.

Table of Contents

List of Abbreviations

ABS   Antilock brake system

CY   Calendar Year

DF   Degrees of Freedom

FARS   Fatality Analysis Reporting System, a census of fatal crashes in the United States since 1975

FMCSA   Federal Motor Carrier Safety Administration

FMVSS   Federal Motor Vehicle Safety Standard

GVWR   Gross vehicle weight rating, specified by the manufacturer, equals the vehicle's curb weight plus maximum recommended loading

IIHS   Insurance Institute for Highway Safety

LTCCS   Large Truck Crash Causation Study

LTV   Light Trucks and Vans, includes pickup trucks, SUVs, minivans, and full-size vans

MY   Model year

NHTSA   National Highway Traffic Safety Administration

PCI   Passenger compartment intrusion

SAS   Statistical analysis software produced by SAS Institute, Inc.

URG   Underride Guard

USDOT   United States Department of Transportation

VIN   Vehicle Identification Number

Background on underride guards

Federal Motor Vehicle Safety Standards (FMVSS) Nos. 223 and 224 require underride guards meeting a strength test on trailers with a GVWR of 10,000 pounds or greater manufactured on or after January 24, 1998. FMVSS No. 224 defines the size requirements for the guards (Figure 1), while FMVSS No. 223 describes strength testing and energy absorption requirements for DOT-compliant guards.[1] The final rule, published in 1996,[2] considers collisions involving passenger vehicles with trucks, trailers, and semitrailers to be an important safety issue, citing 11,551 rear-end collisions that resulted in approximately 423 passenger vehicle occupant fatalities and 5,030 non-fatal injuries.

Figure 1: Configuration requirements for underride guard (FMVSS 224)

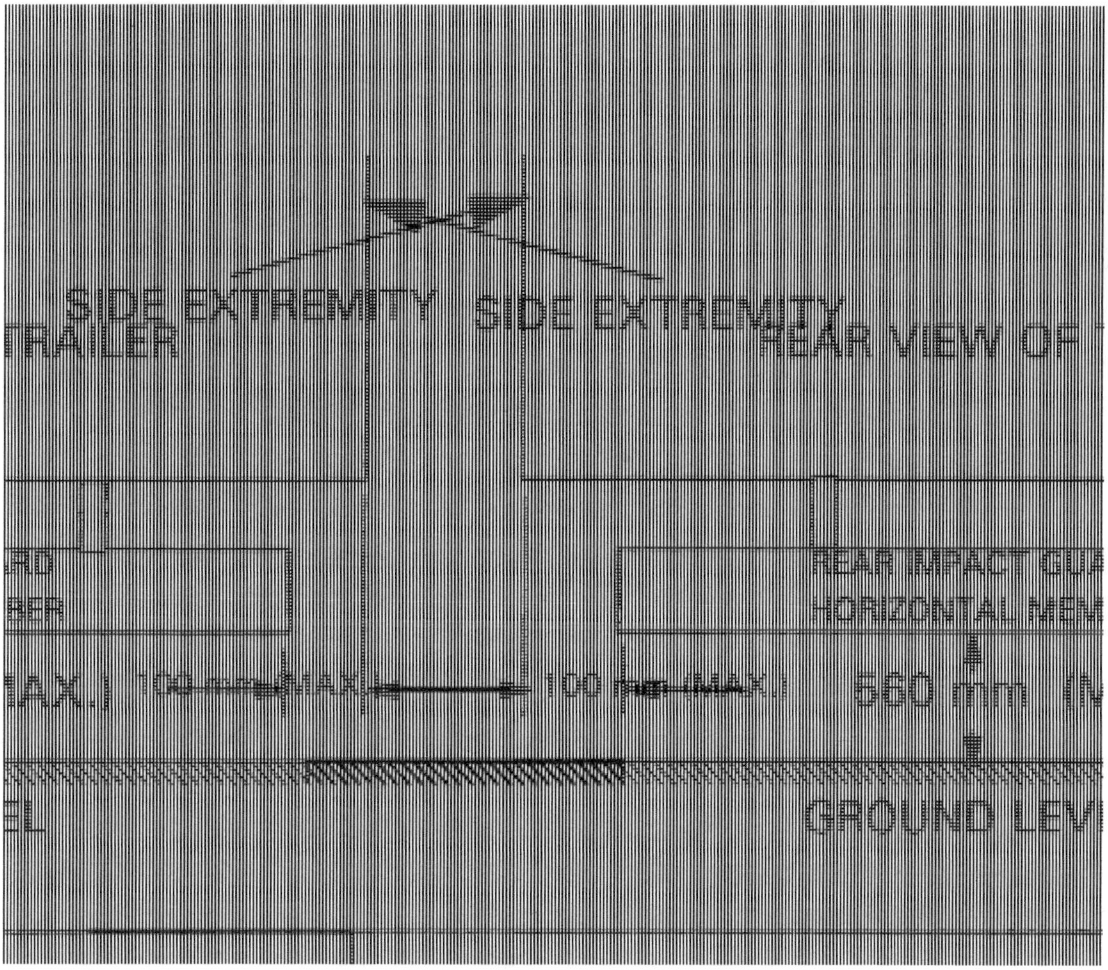

The dimensional requirements of the underride guard are the following (Figure 1):
- Maximum height above ground: 560 mm. (22 inches)
- Maximum distance from side extremities: 100 mm. (4 inches)

---

[1] *Code of Federal Regulations*, Title 49, Parts 571.223 and 571.224
[2] *Federal Register*, Vol. 61, No. 16, page 2004

- Maximum offset from rear plane of trailer: 305 mm. (12 inches)

The energy absorption and strength test requirements are defined at the locations in Figure 2:
- The guard shall resist a force of 50,000 N at points P1 and P2 without deflecting more than 125mm.
- The guard shall resist a force of 100,000 N at point P3 without deflecting more than 125mm.
- The guard shall absorb an energy of 5,560 J within the first 125mm of deflection at each P3 location.

Figure 2: Strength test and energy absorption test locations (FMVSS 223)

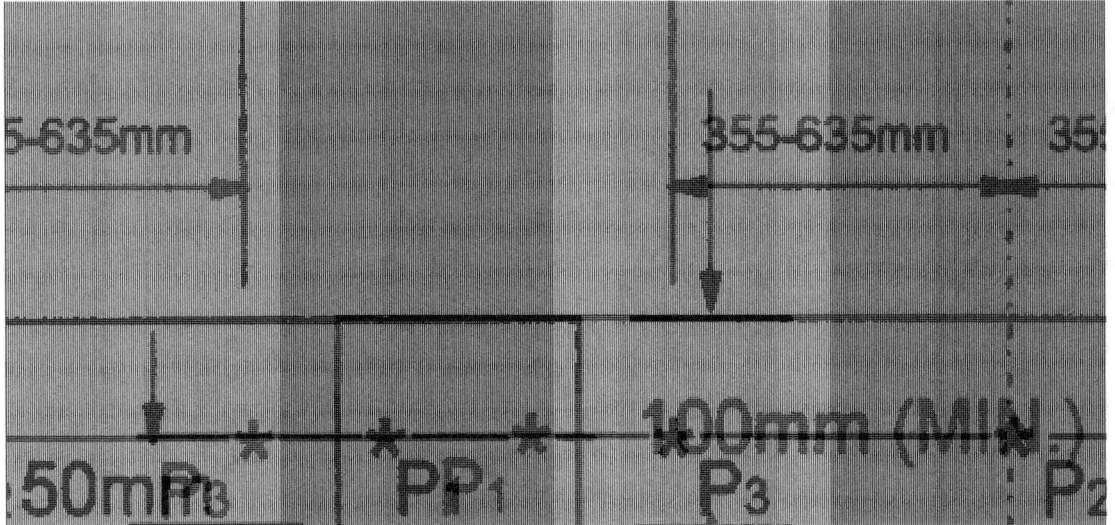

These standards replaced a part of the Federal Motor Carrier Safety Regulations (effective January 1, 1952, to January 25, 1998) that required rear-impact guards but of substantially smaller size and lacking a strength test. In accordance with the Truck Trailer Manufacturers Association's recommended practice (April 1994), some vehicle manufacturers voluntarily installed rear impact guards before 1998. These rear impact guards meet the size requirements of FMVSS No. 224 but it is unknown if they were tested or if they would have met the strength requirements of FMVSS No. 223.

The FMVSS compliant URG is contrasted with the old FHWA style guard in Figure 3. This diagram was provided with a special data collection project in North Carolina, the results of which are discussed later in the report.

Figure 3: Comparison of old FHWA guard and FMVSS-compliant guard (North Carolina special data collection project)

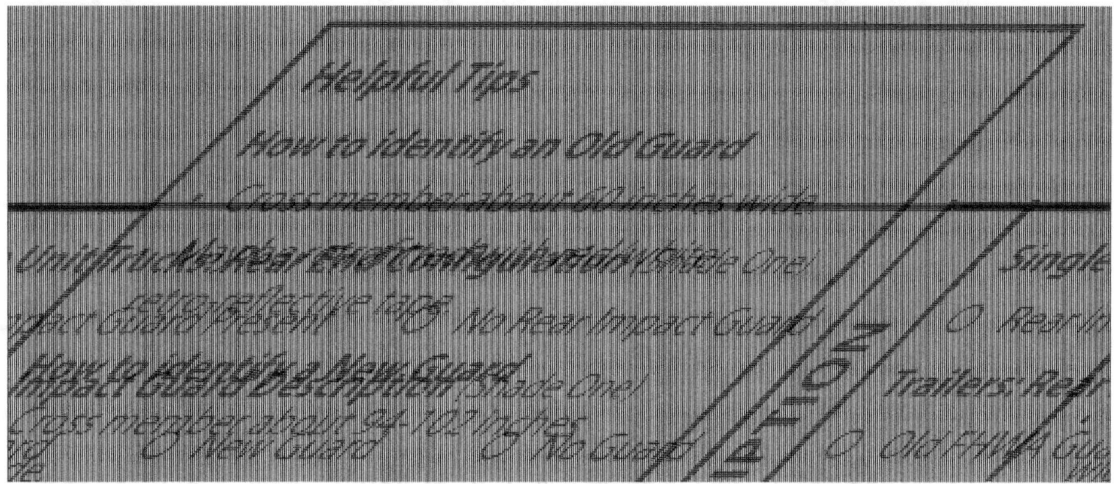

In 1992 and 1993, NHTSA's Vehicle Research & Test Center conducted crash tests of trailers with a guard mounted at 560 mm ground-clearance and attached flush with the trailer. Seven tests were conducted with subcompact cars with low hood profiles (Corsica, Saturn) and masses between 1135 and 1590 kg (2,500 and 3,200 pounds). The tests were conducted at 48 km/h (29 mph). Four of the seven tests resulted in no passenger compartment intrusion when the guard was mounted flush with the rear of the trailer. Two of the seven resulted in PCI due to failure of the guard attachments. One case had PCI by the hood, but the extent was minimal such that the crash test dummy was not contacted.

A Canadian study published in 2000[3] found that an FMVSS 223/224-compliant guard provided good protection of the passenger compartment for an LTV (Windstar mini-van) at 48 km/h (29 mph). This guard was unable to prevent passenger compartment intrusion in a small passenger car (Honda Civic at 56 km/h or 34 mph) or for an average-sized passenger car (Chevrolet Cavalier at 48 km/h or 29mph). A guard with a lower clearance (480 mm instead of 560 mm, refer to Figure 1) allowed successful engagement with both passenger cars at 48 km/h (29 mph) but again failed to prevent compartment intrusion in the Cavalier at 65 km/h (39 mph).

Fatal crashes in the United States

The Fatality Analysis Reporting System is a national census of motor vehicle crashes that result in a fatality to a motor vehicle occupant or a nonmotorist. Crashes involving heavy vehicles, including tractor-trailers, are recorded in FARS, but the model year of the trailer is not provided. FARS can therefore only be used to examine a trend in fatal accidents for each calendar year. As the older trailers with no underride guard or with the older FHWA type are retired or become less used with age, the percentage of the on-road vehicle fleet

---

[3] Boucher, D., & Davis, D. (2000) *Trailer Underride Protection – A Canadian Perspective.* SAE Technical Paper Series, paper 2000-01-3522.

with the FMVSS-compliant underride guard should increase over time. Thus, if the new URG is effective at reducing fatalities, there should be a decrease in fatal crashes where the rear of the trailer is impacted by a passenger vehicle.

The effectiveness of the new URG might be evidenced by a reduction of fatalities in rear-end crashes relative to other types of fatal crashes. Many unknown factors influence the number of crashes per calendar year, such as economic activity and improvements in passenger vehicle front-end safety. However, the many other factors that affect trends in the distributions of crashes involving passenger vehicles and tractor-trailers could mask the effect, if any, of the new URG.

The crashes of interest are the following:
- there are exactly two vehicles
  - VE_FORMS = 2 in the *ACCIDENT* datafile
- one vehicle is a tractor towing exactly one trailer
  - BODY_TYP = 66 (calendar year 1991 to 2008)
  - BODY_TYP = 74 (calendar year 1985 to 1990)
  - TOW_VEH = 1
- the other vehicle is a passenger car or LTV
  - the categories are too numerous to list but can be found in *FARS Analytic Reference Guide*[4]
- the initial point of impact (IMPACT1) is used to identify crashes where a passenger vehicle impacts with its front (IMPACT1 = 11, 12, or 1) into the rear of a tractor-trailer (IMPACT1 = 5, 6, or 7)
- the contrasting group of crashes are those where a passenger vehicle impacts with its front (as above, IMPACT1 = 11, 12, or 1) into some other portion of a tractor-trailer (IMPACT1 = 1, 2, 3, 4, 8, 9, 10, 11, or 12, i.e., all clock points other than 5, 6, and 7)

An important consideration in tractor-trailer impacts is the ambient light condition. According to FMVSS 108, all tractor-trailers manufactured after December 1, 1993 must be equipped with reflective tape. Further, all tractor-trailers, regardless of manufacture date, must have been retrofitted with reflective tape by June 1, 2001. Reflective tape has been shown to be highly effective at reducing side and rear impacts into trailers in dark conditions, especially more severe crashes including those with a fatality[5]. Even if the model year of the trailer were known, it would not be possible to state definitively if it had reflective tape for crashes occurring before June 1, 2001. To account for this in the analysis, crashes are separated according to whether the ambient lighting was daylight (LGT_COND = 1) or not daylight (LGT_COND = 2, 3, 4, or 5).

Crashes that occurred during daylight are the primary condition under which trends will be analyzed. Table 1 shows the number of crashes for three situations:

---

[4] *FARS Analytic Reference Guide1975 to 2008*, page V-7. (2009). DOT HS 811 137. Washington, DC: National Highway Traffic Safety Administration.
[5] Morgan, C. (2001). *The Effectiveness of Retroreflective Tape on Heavy Trailers*, DOT HS 809 222. Washington, DC: National Highway Traffic Safety Administration.

(A)   passenger vehicle impacts with its front into the rear of a tractor-trailer
(B)   passenger vehicle impacts with its front into some other area of a tractor-trailer
(C)   other impacts between a passenger vehicle and a tractor-trailer

Table 1: Number of fatal crashes between a tractor-trailer and a passenger vehicle

	DAYLIGHT				NOT DAYLIGHT				ALL			
CY	(A)	(B)	(C)	(A) ÷ (B+C)	(A)	(B)	(C)	(A) ÷ (B+C)	(A)	(B)	(C)	(A) ÷ (B+C)
1985	74	422	465	0.083	206	361	205	0.364	280	783	670	0.193
1986	82	381	423	0.102	192	336	187	0.367	274	717	610	0.206
1987	88	349	408	0.116	192	364	201	0.340	280	713	609	0.212
1988	86	421	413	0.103	226	340	220	0.404	312	761	633	0.224
1989	106	361	400	0.139	199	366	192	0.357	305	727	592	0.231
1990	99	341	373	0.139	199	323	192	0.386	298	664	565	0.242
1991	79	337	345	0.116	157	284	184	0.335	236	621	529	0.205
1992	92	311	355	0.138	164	322	169	0.334	256	633	524	0.221
1993	99	356	349	0.140	187	305	167	0.396	286	661	516	0.243
1994	107	404	378	0.137	169	321	185	0.334	276	725	563	0.214
1995	99	392	379	0.128	187	293	189	0.388	286	685	568	0.228
1996	126	436	396	0.151	156	347	184	0.294	282	783	580	0.207
1997	117	474	434	0.129	185	346	222	0.326	302	820	656	0.205
1998	129	430	433	0.149	154	375	191	0.272	283	805	624	0.198
1999	142	430	463	0.159	136	353	211	0.241	278	783	674	0.191
2000	122	446	423	0.140	166	338	196	0.311	288	784	619	0.205
2001	127	472	395	0.146	118	297	179	0.248	245	769	574	0.182
2002	123	432	429	0.143	138	310	194	0.274	261	742	623	0.191
2003	131	410	386	0.165	151	291	222	0.294	282	701	608	0.215
2004	159	438	418	0.186	145	269	220	0.297	304	707	638	0.226
2005	142	429	402	0.171	152	289	190	0.317	294	718	592	0.224
2006	132	370	386	0.175	145	284	186	0.309	277	654	572	0.226
2007	136	412	345	0.180	119	246	184	0.277	255	658	529	0.215
2008	96	314	330	0.149	120	227	157	0.313	216	541	487	0.210

The situation (A) is where the presence of the underride guard should show effectiveness. The situations (B) and (C) are where the presence of the underride guard should be irrelevant. The ratio of (A) over (B+C) is referred to as the *odds* (Figure 4). The odds should decrease, subject to controlling for ambient light condition. In daylight, there should be a decrease in rear-impacting fatalities compared to other fatalities. Clearly, this is not the case – the odds increases from around 0.10 in the earliest years to nearly 0.20 in the latest years. In the non-daylight condition, however, what is assumed to be the effect of conspicuity tape is confirmed – the number of crashes in each column decreases, to an extent that the odds remains approximately constant.

The effect of conspicuity tape is not explicitly in Table 1 because the (B) column includes both frontal and side impacts. With some re-arrangement (not shown), the (B) crashes can be divided to crudely estimate the effect of conspicuity tape – [(A) + (B side)] ÷ [(B front) + (C)]. For the earliest calendar years, 1985 to 1989, when very few trailers would have had conspicuity tape or an underride guard, there were 1,598 impacts into the rear or side of a tractor-trailer in non-daylight and 2,189 impacts into the front of a tractor-trailer or other crashes between a passenger vehicle and tractor-trailer in non-daylight (1,598 ÷ 2,189 = 0.730). For the most recent calendar years 2004 to 2008, when all trailers should have had conspicuity tape and many newer trailers with an FMVSS-compliant underride guard would have been in use, there were 1,173 front or side impacts, compared to 1,760 front impacts and other crashes. The ratio is 0.666, and it is 9 percent lower than that from 1985 to 1989. The analysis of conspicuity tape reported a reduction of 29 percent of collisions into the side or rear of trailers, using more rigorous methodology.[5]

A potentially confounding factor is the type of passenger vehicle – i.e., a passenger car or an LTV. Passenger cars, since they are typically lower to the ground, could be more apt to experience under-ride, regardless of the portion of the tractor-trailer impacted. Since the proportion of the vehicle fleet that is passenger car versus LTV varies by calendar year, this factor could undermine what is shown in Figure 4.

Table 2 shows the portion of Table 1 where the passenger vehicle is a **passenger car**. For both *daylight* and *not daylight*, the trends in the ratio (A) ÷ (B+C) are comparable to those for all passenger vehicles.

Table 2: Number of fatal crashes between a tractor-trailer and a passenger car

	DAYLIGHT				NOT DAYLIGHT			
CY	(A)	(B)	(C)	(A) ÷ (B+C)	(A)	(B)	(C)	(A) ÷ (B+C)
1985	47	316	368	0.069	155	264	166	0.360
1986	54	299	333	0.085	142	259	154	0.344
1987	70	265	315	0.121	147	255	159	0.355
1988	59	323	328	0.091	170	249	167	0.409
1989	64	235	307	0.118	135	270	149	0.322
1990	69	236	289	0.131	127	221	142	0.350
1991	52	220	249	0.111	102	192	131	0.316
1992	62	220	268	0.127	112	214	125	0.330
1993	63	247	257	0.125	125	198	110	0.406
1994	66	273	264	0.123	109	203	136	0.322
1995	55	255	268	0.105	133	184	136	0.416
1996	75	274	278	0.136	94	229	127	0.264
1997	65	285	296	0.112	121	220	150	0.327
1998	76	266	288	0.137	91	214	126	0.268
1999	84	263	324	0.143	83	195	139	0.249
2000	70	261	282	0.129	95	200	132	0.286
2001	63	280	251	0.119	73	179	106	0.256
2002	66	245	276	0.127	83	171	124	0.281
2003	61	233	241	0.129	84	167	143	0.271
2004	83	248	263	0.162	83	148	130	0.299
2005	68	244	231	0.143	83	151	112	0.316
2006	64	190	240	0.149	82	156	102	0.318
2007	67	216	204	0.160	68	125	112	0.287
2008	45	166	189	0.127	66	116	93	0.316

Table 3 shows the crashes from Table 1 where the passenger vehicle is a **light truck or van**. Again, the trends in the target ratio (A) ÷ (B+C) are similar to those for passenger cars. Thus, it does not appear that the classification of passenger car versus LTV is undermining the trend in Table 1.

Table 3: Number of fatal crashes between a tractor-trailer and a LTV

	DAYLIGHT				NOT DAYLIGHT			
CY	(A)	(B)	(C)	(A) ÷ (B+C)	(A)	(B)	(C)	(A) ÷ (B+C)
1985	27	106	97	0.133	51	97	39	0.375
1986	28	82	90	0.163	50	77	33	0.455
1987	18	84	93	0.102	45	109	42	0.298
1988	27	98	85	0.148	56	91	53	0.389
1989	42	126	93	0.192	64	96	43	0.460
1990	30	105	84	0.159	72	102	50	0.474
1991	27	117	96	0.127	55	92	53	0.379
1992	30	91	87	0.169	52	108	44	0.342
1993	36	109	92	0.179	62	107	57	0.378
1994	41	131	114	0.167	60	118	49	0.359
1995	44	137	111	0.177	54	109	53	0.333
1996	51	162	118	0.182	62	118	57	0.354
1997	52	189	138	0.159	64	126	72	0.323
1998	53	164	145	0.172	63	161	65	0.279
1999	58	167	139	0.190	53	158	72	0.230
2000	52	185	141	0.160	71	138	64	0.351
2001	64	192	144	0.190	45	118	73	0.236
2002	57	187	153	0.168	55	139	70	0.263
2003	70	177	145	0.217	67	124	79	0.330
2004	76	190	155	0.220	62	121	90	0.294
2005	74	185	171	0.208	69	138	78	0.319
2006	68	180	146	0.209	63	128	84	0.297
2007	69	196	141	0.205	51	121	72	0.264
2008	51	148	141	0.176	54	111	64	0.309

Table 2 and Table 3 are depicted graphically in Figure 4. The points are the annual values of the odds – from the column (A) ÷ (B+C). The trend line is a simple linear regression on the odds for each calendar year.

8

Figure 4: Odds of rear-end fatal crashes to other fatal crashes, by ambient lighting condition

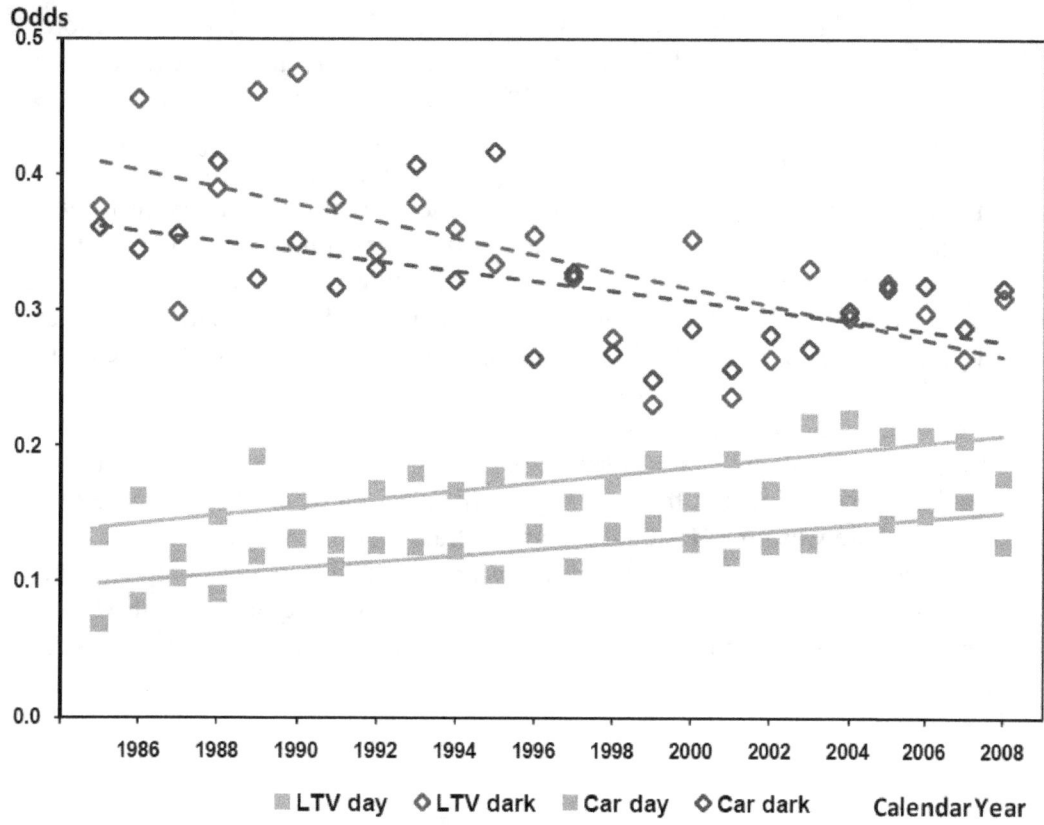

Several observations can be made:
  – The difference between daylight and darkness becomes less important in more
    recent calendar years.
  – The difference between daylight and darkness is more exaggerated for cars than
    for LTVs.
  – During daylight, when the effectiveness of the underride guard is not confounded
    by the presence of conspicuity tape, the trend is not in the expected direction. That
    is, the ratio of fatal crashes into the rear of a tractor-trailer should decrease
    relative to fatal crashes into other portions of a tractor-trailer.

The results presented in Table 1, Table 2, and Table 3, which are summarized in Figure 4,
fail to reveal that the underride guard is effective at preventing fatal crashes where a
passenger vehicle impacts the rear of a tractor-trailer. The lighting condition is crucial, to
control for the presence of conspicuity tape on the trailer.

The following figures are analogous to Figure 4 for several other potentially confounding
conditions. The data for these figures are restricted to *daylight* crashes.
  – Type of road: *interstate* versus *non-interstate* (Figure 5). The type of road can
    represent differences in both the prevailing road speeds and the type of
    surrounding traffic.

- Type of locality: *urban interstate* versus *urban non-interstate* (Figure 6). Nearly all rural crashes occur on interstates, and these crashes on rural interstates are the largest contributor to the total number of crashes. Crashes in urban areas more commonly occur on interstates but to a lesser degree than in rural localities.
- Presence of tractor anti-lock brakes (ABS): *tractor MY 96-* versus *tractor MY 98+* (Figure 7). FMVSS No. 121, Air Brake Systems, mandates antilock braking systems on virtually all new air-braked vehicles with a GVWR of 10,000 pounds or greater. ABS is required on tractors manufactured on or after March 1, 1997 and air-braked semi-trailers and single-unit trucks manufactured on or after March 1, 1998. This means that some tractors manufactured with the model year 1997 may not have ABS and some with model years 1996 and earlier might have voluntary ABS installations. The ABS might be a secondary effect that could alter the types of crashes tractor-trailers are involved in.
- Age of tractor at the time of the crash: *tractor age ≤ 5* versus *tractor age ≥ 6* (Figure 8). The model year as used above to investigate ABS equipment is similar to tractor age, but the full range of data cannot be used to estimate a trend because there was no ABS requirement for tractors prior to calendar year 1997. Using age explicitly, all calendar years can be compared. If viewed in tandem with ABS, there are several combinations, such as old tractors without ABS versus new tractors without ABS or old tractors without ABS versus new tractors with ABS.

Figure 5: Odds of rear-end fatal crashes to other fatal crashes in daylight, by type of road

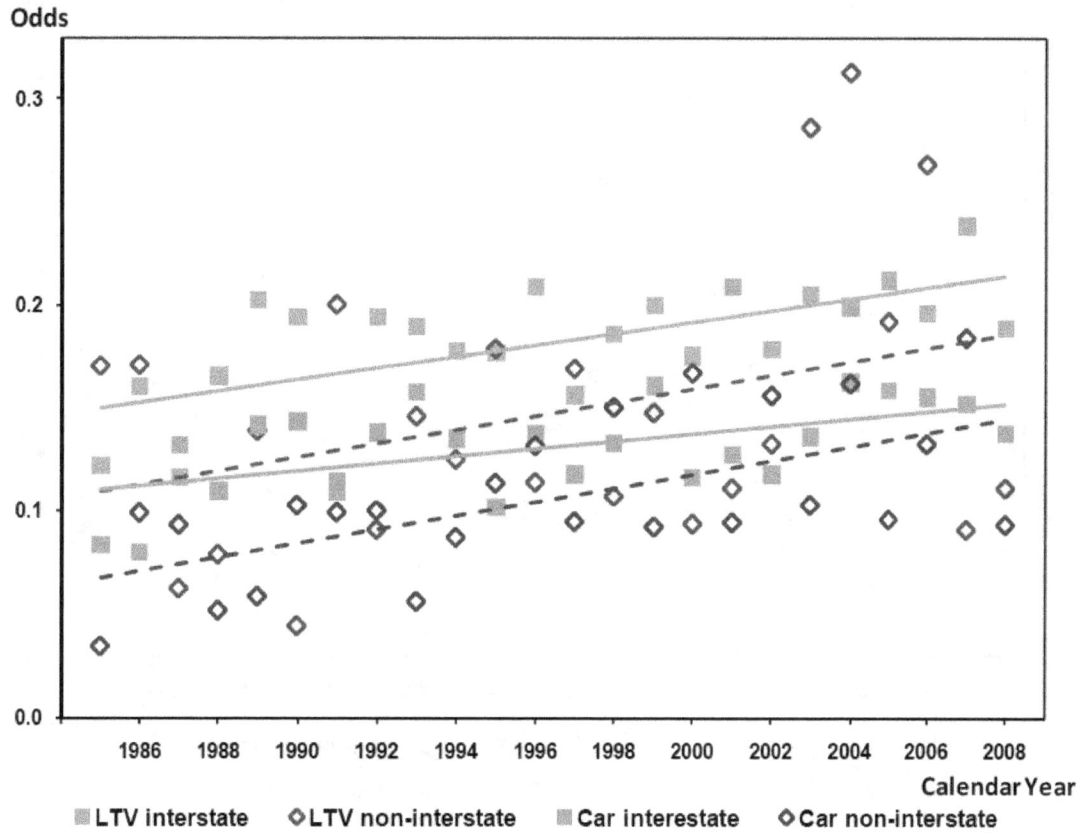

Figure 6: Odds of rear-end fatal crashes to other fatal crashes in daylight in urban areas, by road type

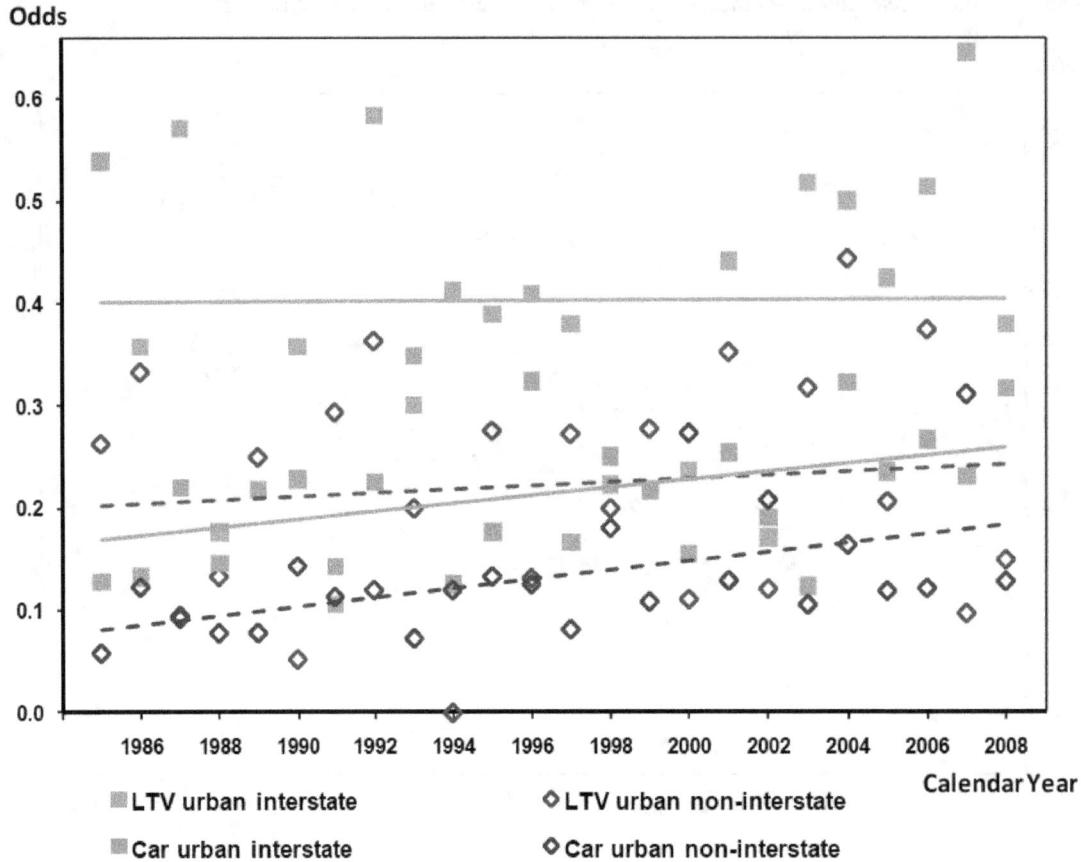

Figure 7: Odds of rear-end fatal crashes to other fatal crashes in daylight, by tractor model year

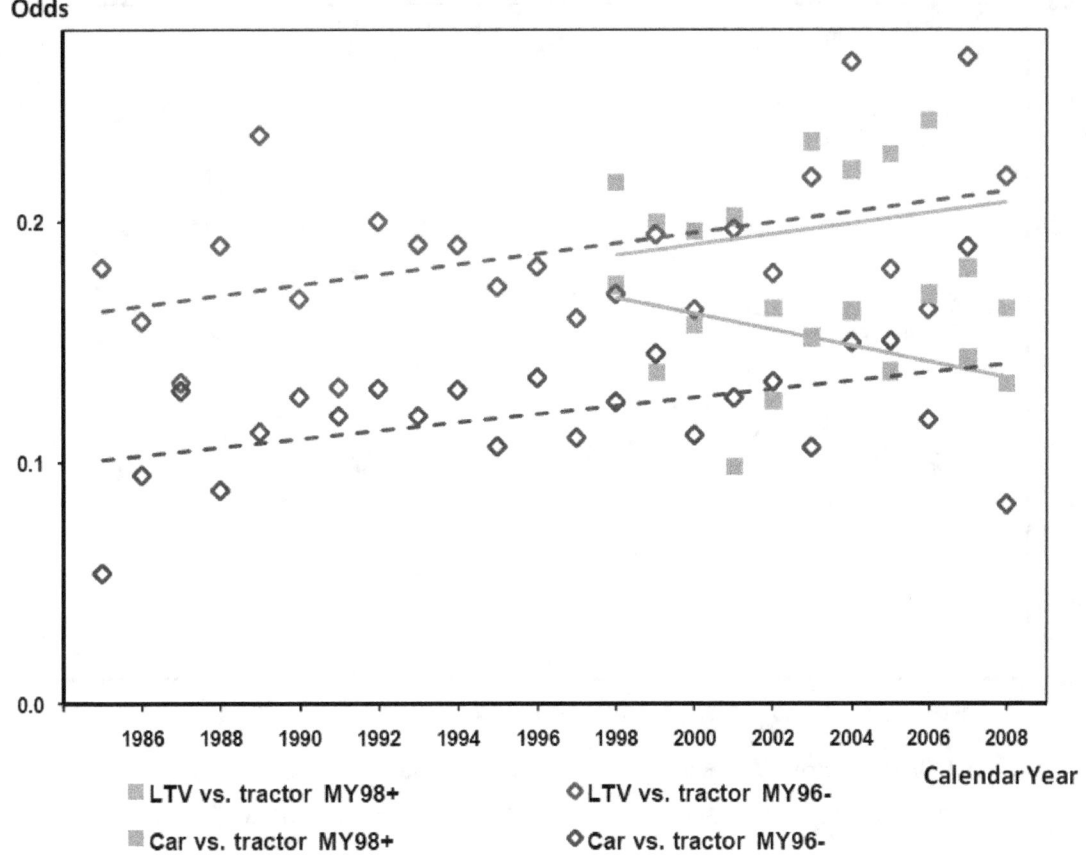

Figure 8: Odds of rear-end fatal crashes to other fatal crashes in daylight, by age of tractor

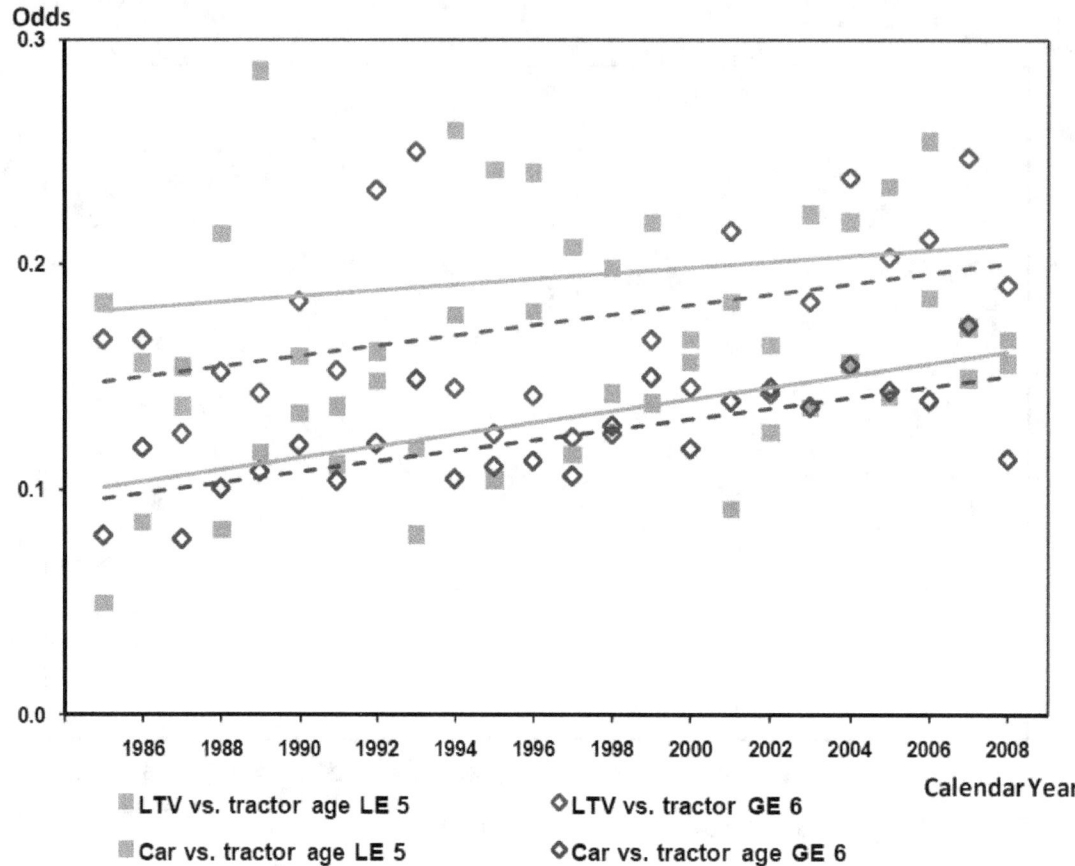

These figures display generally the same basic trend as the overall *daylight* crashes in Figure 4. That is, tractor-trailers are more apt to be struck in the rear compared to being involved in some other accident, in two-vehicle crashes with a fatality in a passenger vehicle. (The only downward trend line is for cars versus tractors with ABS in Figure 7, but there are very few of these in the calendar years up to 2001.)

The analysis of year-to-year trends in overall fatalities does not indicate a significant reduction in rear-impact fatalities over time, relative to other types of fatal collisions between passenger vehicles and tractor-trailers. Because FARS does not provide the model year of the trailer, this is the only possible analysis with FARS data exclusively.

The data from FARS can be supplemented with those from NHTSA's General Estimates System (GES), a nationally-representative probability sample of all crashes. FARS and GES contain variable definitions that are similar enough for comparison back to calendar year 1991. Table 4 shows the number of crashes from these two sources, restricted to daylight crashes (for FARS, these are the same as in the daylight part of Table 1). At right, the rate of fatalities per 1,000 crashes is the measure of interest.

14

Table 4: Number of fatalities compared to total number of crashes, exactly one passenger vehicle and exactly one tractor-trailer

CY	Fatalities (FARS)		All crashes (GES)		Fatalities per 1000 crashes	
	Rear	Other	Rear	Other	Rear	Other
1991	79	682	11,232	10,433	7.0	65.4
1992	92	666	9,714	11,199	9.5	59.5
1993	99	705	9,857	11,180	10.0	63.1
1994	107	782	13,674	12,569	7.8	62.2
1995	99	771	7,810	11,139	12.7	69.2
1996	126	832	12,513	13,681	10.1	60.8
1997	117	908	13,953	13,008	8.4	69.8
1998	129	863	10,364	11,869	12.4	72.7
1999	142	893	14,835	15,157	9.6	58.9
2000	122	869	10,988	14,295	11.1	60.8
2001	127	867	10,133	15,417	12.5	56.2
2002	123	861	9,114	11,850	13.5	72.7
2003	131	796	9,155	11,270	14.3	70.6
2004	159	856	12,252	12,396	13.0	69.1
2005	142	831	12,517	17,827	11.3	46.6
2006	132	756	8,780	11,905	15.0	63.5
2007	136	757	11,364	12,540	12.0	60.4
2008	96	644	9,599	11,272	10.0	57.1

In Figure 9, the fatality rate is compared for passenger vehicles impacting the rear of a tractor trailer versus other impacts between a passenger vehicle and tractor trailer. The two are mapped to different axes because the magnitudes vary greatly (Rear on the left-hand scale, Other on the right-hand scale). The fatality rate in rear impacts increases slightly over the time frame 1991 to 2008, from around 9 fatalities per 1,000 crashes to 14 fatalities per 1,000 crashes. By comparison, the fatality rate in other impacts between a tractor trailer and passenger vehicle is relatively constant or slightly decreasing, with a rate just above 60 fatalities per 1,000 crashes.

Figure 9: Fatalities per 1,000 crashes for rear impacts (left axis) compared to other impacts (right axis)

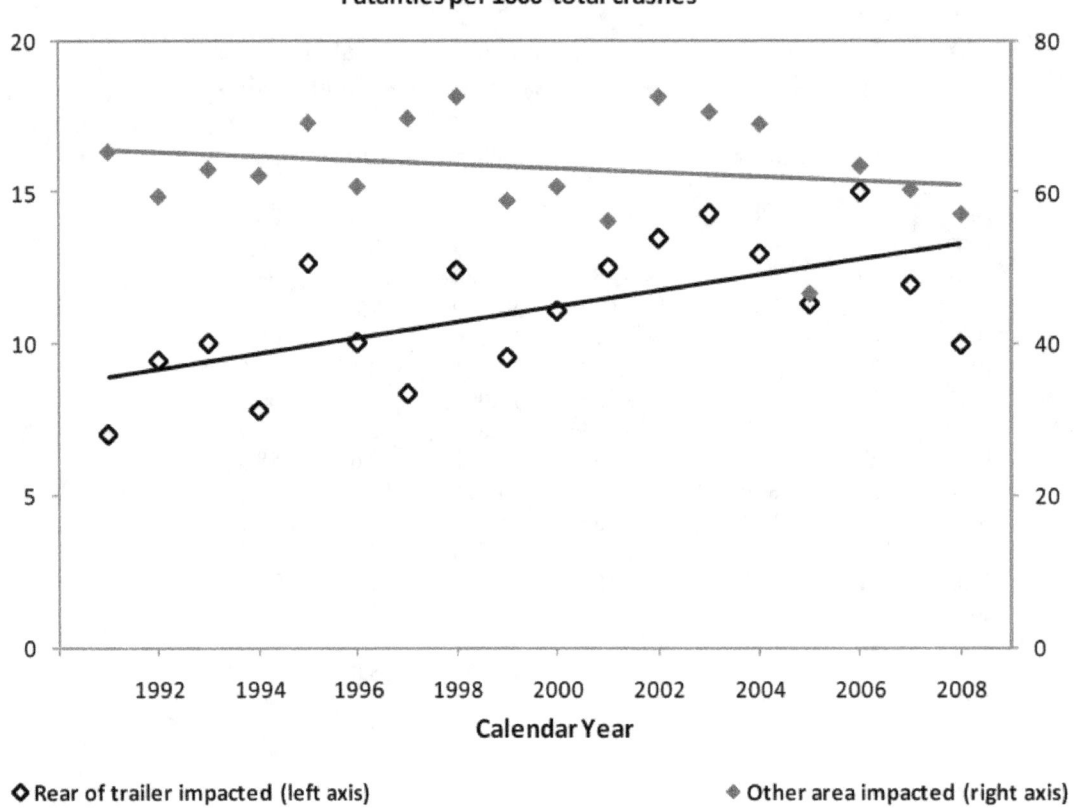

The analysis shows that the number of fatal crashes in which a passenger vehicle rear-impacts a tractor-trailer continues to be high.

Analysis of fatal and nonfatal crashes in two States

Florida – State Data System

Data from Florida are available through NHTSA's State Data System. Data from calendar years 1989 to 2006 are available. Florida is the only State that records information about the trailer unit relevant to this analysis – the type of trailer, its model year, and its VIN. Other States in NHTSA's SDS record the presence of a trailer or the number of trailers towed by a tractor.

Tractor-trailer combinations are identified by the type of vehicle (VEH_TYPE) and type of trailer (TRL_TYPE). The special vehicle code (SPEC_VEH) is included to exclude peculiarities such as boat trailers being hauled by a heavy truck. The variable definitions are as follows:

VEH_TYPE	05	Heavy truck
	06	Truck-tractor
SPEC_VEH	03	Commercial cargo
TRL_TYPE	01	Single semi-trailer (1993 and later)
	26	Trailer (1989 to 1992)

Passenger vehicles are likewise identified by the vehicle type element, listed below. The value listed as "04 Medium truck (4 rear tires)" should generally be composed of larger passenger pickups, based on reviewing the VIN. The number of vehicles is small enough that it cannot have a large influence (around 4% of all passenger vehicles in this analysis for calendar years 2002 and later have VEH_TYPE = "04"). A review of reporting inconsistencies in Florida was conducted by The University of Michigan's Transportation Research Institute.[6]

VEH_TYPE	01	Car	
	02	Passenger van	
	03	Recreational	(2001 and earlier)
	03	Pickup/light truck (2 rear tires)	(2002 and later)
	04	Truck—light pickup	(2001 and earlier)
	04	Medium truck (4 rear tires)	(2002 and later)

As noted in the FARS analysis, ambient light condition is a pivotal factor that must be accounted for. There are a small number of crashes (< 1%) where the lighting condition is undefined, and these are considered to be daylight if the time of the crash is between 8:00am and 8:00pm.

LIGHT	01	Daylight
	02	Dusk
	03	Dawn
	04	Darkness – street lights present
	05	Darkness – no street lights

The impact zones identify the rear, front, and side portions of the vehicle. Additionally, the trailer can be identified. In order for a crash to be considered into the rear of a trailer, the vehicle event is also used.

IMPACT	6, 7, 8, 9, 10	Rear portions of vehicle
	1, 2, 3, 13, 14	Front portions of vehicle
	4, 5, 11, 12	Side portions of the vehicle

IMPACT	21	Trailer (2002 and later)
	22	Trailer (2001 and earlier)
and		
EVENT	1	Rear-end

The trailer model year (TRL_YR) is used to identify the type of underride guard. There are three eras:

---

[6] Blower, D., & Matteson, A. (2004). *Evaluation of Florida Crash Data Reported to MCMIS Crash File*, Report No. UMTRI-2004-41. Washington, DC: Federal Motor Carrier Safety Administration.

TRL_YR   1980 to 1993   Smaller guard – narrower and set further inwards
                   1994 to 1997   Voluntary installation of larger guards based on TTMA recommendation
                   1998 and newer   NHTSA mandated guard (Figure 1)

The injury status INJ of the driver in the passenger vehicle defines the severity of the crash. There are five levels, shown below. The equivalent KABCN scale is given in parenthesis for reference[7].

      INJ   1   (N) None
             2   (C) Possible injury
             3   (B) Non-incapacitating injury
             4   (A) Incapacitating injury
             5   (K) Fatal injury

Table 5 shows the number of crashes where a passenger vehicle impacts a tractor-trailer. These crashes include crashes under all ambient light conditions. Later, when performing the effectiveness calculations (Table 6 and Table 7), the crashes that happened before 2002 will be restricted to daylight. "Rear impacts" are defined in terms of the impact zone and include cases where the impact zone is the trailer if the event is a rear-end collision. "Other impacts" are either into the front or side based on the impact zone or into the trailer if the event is not a rear-end collision. The injury status in the passenger vehicle is listed according to the KABCN scale.

Table 5: Number of rear impacts and other impacts between a passenger vehicle and tractor-trailer, by injury outcome for driver of passenger vehicle

		Rear impacts					Other impacts				
	Trailer MY	N	C	B	A	K	N	C	B	A	K
CY 1993 to 2001	1980 to 1993	300	112	178	113	53	600	193	261	164	83
	1994 to 1997	124	50	67	50	10	306	75	103	70	25
	1998 & new er	62	27	33	18	2	194	42	66	35	16
CY 2002 to 2005	Trailer MY	N	C	B	A	K	N	C	B	A	K
	1980 to 1993	269	63	62	42	15	1494	311	246	112	48
	1994 to 1997	267	49	74	42	13	1507	292	196	111	27
	1998 & new er	483	89	125	68	13	2831	510	360	192	63
CY 2006 & 2007	Trailer MY	N	C	B	A	K	N	C	B	A	K
	1980 to 1993	49	30	15	12	2	187	85	65	31	11
	1994 to 1997	53	26	18	14	5	245	124	85	42	11
	1998 & new er	160	61	70	30	9	671	359	246	104	25

The composition of the data varies in terms of the number of relevant crashes that can be identified and the distribution of injury severity. Figure 10 shows the number of crashes included in Table 5, grouped by calendar year. Figure 11 converts these counts to a percentage basis for each calendar year. There are four distinct regions in the graphics:

---

[7] This scale is also referred to as KABCO. The value O is sometimes called "Other" or "Property Damage Only," depending on the data source. The term U ("Uninjured") is encountered as well.

- 1989 to 1992 – These years are excluded from the analysis because there are only the oldest model year trailers (MY 1980 to 1993), thus allowing no comparison to the trailers with FMVSS-compliant underride guards. There is also less certainty in the identification of typical tractor-trailers because the trailer type variable (TRL_TYPE) lists only one value "trailer" as opposed to "single semi-trailer".
- 1993 to 2001 – Figure 11 shows a year-to-year increase in the percentage of non-injury crashes. These property-damage-only crashes may become more common in later calendar years because the reporting criterion is fixed at $500. Most simply, this could be due to inflation in the economy as a whole. It may also reflect advances in vehicle technology, such as modifications to the headlamp housings, which are more expensive to repair on newer vehicles.
- 2002 to 2005 – The basic format of the data changed from 2001 to 2002. It appears there was better recording of the trailer model year or VIN, allowing a greater number of vehicles to be identified for this study. The proportion of non-injury crashes is also higher.
- 2006 and 2007 – The proportion of non-injury crashes is comparable to the earlier 1993 to 2001 range. The number of identifiable crashes falls to about half of that from 2002 to 2005, though still much higher than the number from 2001 and earlier.

Figure 10: Number of crashes between a tractor-trailer and passenger vehicle, by calendar year and injury outcome

Crashes per Calendar Year

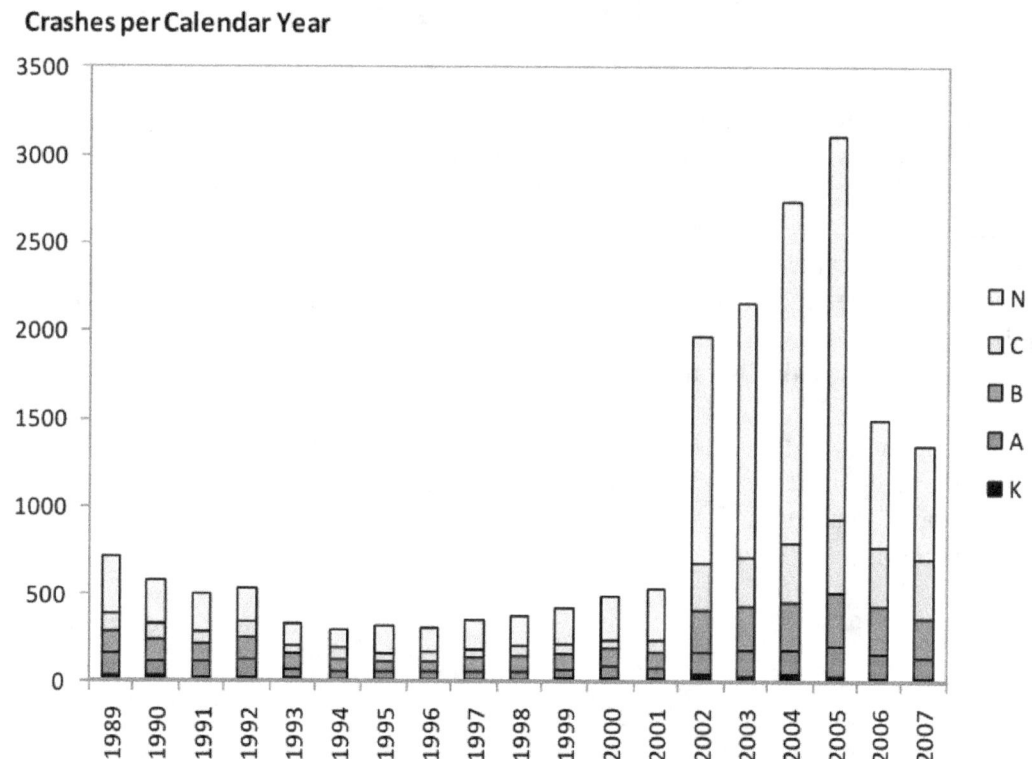

Figure 11: Relative number of crashes between a tractor-trailer and passenger vehicle, by calendar year and injury outcome

Distribution of crashes per Calendar Year

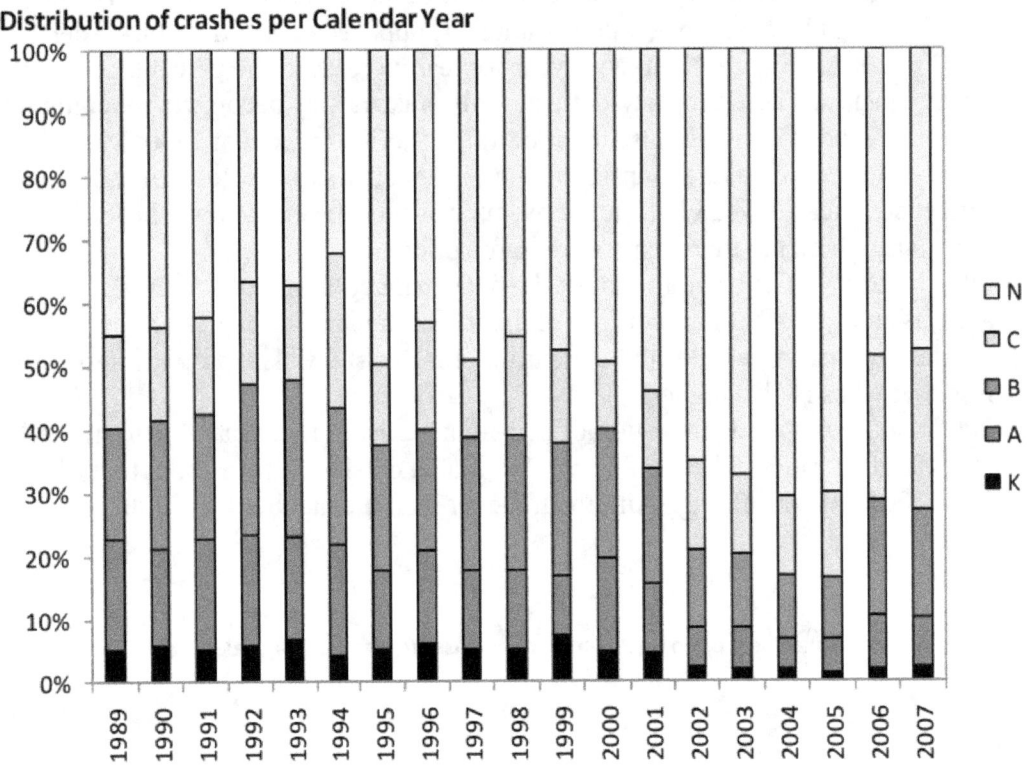

Categorical data modeling accounts for the differing proportion of crashes across the calendar year groups and arrives at an overall estimate of the effectiveness of the underride guard. The method is a generalization of a 2×2 table, allowing the incorporation of multiple categories in more than two dimensions. Similar to logistic regression, categorical data modeling estimates the effect of each value of the independent variable, and of combinations of independent variables (interaction terms), upon the dependent variable (fatal or nonfatal). The *proc catmod* procedure of SAS is used to perform these calculations.

Table 6 shows the overall model results from the *catmod* procedure. The Likelihood Ratio has a non-significant *chi-square* (p-value > 0.05). In this test, the null hypothesis is that the model is an accurate fit to the observed data. Therefore, since the null hypothesis is not rejected, the model matches the observed data within an acceptable margin of error.

Table 6: CATMOD summary, fatality crashes

TERM	DF	CHI-SQ	p-value
Intercept	1	1511.75	< 0.0001
Cohort	2	5.94	0.051
Target	1	0.09	0.770
Cohort*Target	2	2.88	0.236
Urg	1	3.18	0.075
Target*Urg	1	0.88	0.347
Cohort*Urg	2	1.00	0.607
Likelihood Ratio	2	3.92	0.141

The three main effects are the following:
- *Cohort* – refers to the three calendar year ranges (1993 to 2001; 2002 to 2005; 2006 and 2007) and models the extent to which the odds ratio of fatalities to non-fatalities in reported crashes has varied over the years.
- *Target* – refers to the region of the tractor-trailer impacted by the passenger vehicle (Rear versus Other) and models the extent to which fatality risk is higher (or lower) in rear impacts than in other impacts.
- *Urg* – refers to the underride guard on the trailer, based on model year (1980 to 1993 = None or Old FHWA type; 1998 and newer = FMVSS 223 and 224 compliant guard) and models the extent to which fatality risk is higher (or lower) in vehicles with the newer trailer. This term is a measure of the change in overall fatality rates for both rear impacts and other impacts. (The intermediate model years 1994-1997 are not included in the analysis because of uncertainty about the extent of voluntary compliance.)

The two-way interactions can be interpreted as follows:
- *Target * Urg* – this is the term that estimates the effect of the underride guard at reducing rear impact fatalities relative to fatalities in other impacts
- *Cohort * Target* – accounts for variation in the "other impacts" crashes (which is used as a control group for the "rear impacts" crashes)
- *Cohort * Urg* – accounts for the extent that the difference in overall fatality rates varies across the cohorts for the two URG categories

The three-way interaction *Cohort * Target * Urg* was excluded from the analysis. Its inclusion would imply that the effectiveness of the underride guard varies across the cohorts.

The observed data (Table 5) and model-predicted number of crashes are shown in Table 7. The values marked OR(1) and OR(2) are the odds ratios for each adjacent 2×2 table. The calculation OR(1) ÷ OR(2) is the estimate of the effectiveness. The model controls for the differences in number of exposure crashes (rear impacts) relative to a control group of other impacts. Although the counts are very similar between the observed and predicted tables, the slight adjustments invoked by the model arrive at a constant effectiveness for the three cohorts. The final estimate of the effectiveness is 27.1 percent. However, this value is not statistically significant because the *Target * Urg* interaction has a chi-square of 0.88 (*p*-value > 0.10 in Table 6).

Table 7: Observed and Predicted crash frequencies, fatality crashes versus all others

		Observed				Predicted		
CY 1993 to 2001	Rear Impacts	K	ABCN			K	ABCN	
	1980 to 1993	16	423			14	425	
	1998 & newer	1	100			3	98	
				0.264				0.751 OR(1)
	Other Impacts							
	1980 to 1993	35	869			37	867	
	1998 & newer	12	240			10	242	
				1.241				1.031 OR(2)
					0.213			0.729 OR(1) ÷ OR(2) (Effectiveness)
CY 2002 to 2005	Rear Impacts	K	ABCN			K	ABCN	
	1980 to 1993	15	436			15	436	
	1998 & newer	13	765			13	765	
				0.494				0.524 OR(1)
	Other Impacts							
	1980 to 1993	48	2163			48	2163	
	1998 & newer	63	3893			63	3893	
				0.729				0.718 OR(2)
					0.677			0.729 OR(1) ÷ OR(2) (Effectiveness)
CY 2006 & 2007	Rear Impacts	K	ABCN			K	ABCN	
	1980 to 1993	2	106			4	104	
	1998 & newer	9	321			7	323	
				1.486				0.581 OR(1)
	Other Impacts							
	1980 to 1993	11	368			9	370	
	1998 & newer	25	1380			27	1378	
				0.606				0.797 OR(2)
					2.452			0.729 OR(1) ÷ OR(2) (Effectiveness)

The next analysis aims to find a reduction in serious injuries and fatalities (injury types K and A). The model is a good fit for the data (Likelihood ratio = 0.27). The final estimate of the effectiveness at reducing fatalities and serious injuries is 6.5 percent. Again, this estimate is not statistically significant because the chi-square for the *Target * Urg* term is 0.16 (*p*-value > 0.10).

Table 8: CATMOD summary, fatality and serious injury crashes

TERM	DF	CHI-SQ	p-value
Intercept	1	1961.51	< 0.0001
Cohort	2	22.45	< 0.0001
Target	1	12.82	< 0.01
Cohort*Target	2	7.83	0.020
Urg	1	2.16	0.142
Target*Urg	1	0.16	0.685
Cohort*Urg	2	0.44	0.801
Likelihood Ratio	2	0.27	0.874

Table 9: Observed and Predicted crash frequencies, fatality and serious injury crashes versus all others

		Observed				Predicted		
		KA	BCN			KA	BCN	
CY 1993 to 2001	Rear Impacts							
	1980 to 1993	64	375			63	376	
	1998 & newer	13	88			14	87	
				0.866				0.920 OR(1)
	Other Impacts							
	1980 to 1993	121	783			122	782	
	1998 & newer	34	218			33	219	
				1.009				0.984 OR(2)
					0.858			0.935 OR(1) ÷ OR(2) (Effectiveness)
CY 2002 to 2005	Rear Impacts							
	1980 to 1993	57	394			56	395	
	1998 & newer	81	697			82	696	
				0.803				0.820 OR(1)
	Other Impacts							
	1980 to 1993	160	2051			161	2050	
	1998 & newer	255	3701			254	3702	
				0.883				0.877 OR(2)
					0.910			0.935 OR(1) ÷ OR(2) (Effectiveness)
CY 2006 & 2007	Rear Impacts							
	1980 to 1993	14	94			15	93	
	1998 & newer	39	291			38	292	
				0.900				0.791 OR(1)
	Other Impacts							
	1980 to 1993	42	337			41	338	
	1998 & newer	129	1276			130	1275	
				0.811				0.846 OR(2)
					1.109			0.935 OR(1) ÷ OR(2) (Effectiveness)

Similar to the presentation in the analysis of fatal crashes (Figure 5 - Figure 8), several other possibly influential variables were introduced into the categorical data model – type of roadway, whether the passenger vehicle was a car or an LTV, whether the passenger vehicle driver was restrained, and the ABS equipment of the tractor based on the model year. Initially, a model was fit with all three-way interactions. The three-way interactions with the lowest chi-square values were removed from the model until only those with a statistically-significant contribution remained. The process was then repeated on the two-way interactions[8]. In the end, there was no appreciable change in the model fit nor the predicted frequencies (Table 9). Therefore, none of these other external variables are presented in the final estimate of URG effectiveness.

North Carolina – Special data collection

The basic analysis for Florida is repeated using a special data collection project from North Carolina. From 2005 to 2007, the State Highway Patrol filed a supplementary crash report for crashes involving a heavy truck, in addition to the State's standard crash report. The additional information includes characteristics of the trailer that are relevant to the current analysis but do not appear on the standard form.

Table 10 shows the number of crashes involving one passenger vehicle and one tractor-trailer, according to whether the passenger vehicle impacted the rear of the tractor-trailer or some other portion.

Table 10: Number of passenger vehicle crashes into the rear versus other portions of a tractor-trailer, according to injury status and model year of the trailer

Rear impacts

Trailer MY	N	C	B	A	K
1980 to 1993	47	16	8	3	3
1994 to 1997	57	10	11	1	2
1998 & newer	144	27	26	3	2

Other impacts

Trailer MY	N	C	B	A	K
1980 to 1993	291	72	35	19	6
1994 to 1997	310	90	30	8	12
1998 & newer	904	222	94	28	26

The data in Table 10 were subjected to a categorical data analysis. Because there is no calendar year cohort to control for, the model is simpler but the interpretation is different. The two-way interaction (*Target * Urg*) is excluded so that the Likelihood Ratio can be estimated. This term becomes the item of interest because it accounts for model variance that would belong to the two-way interaction. In the Florida analysis (Table 6 and Table 8), the Likelihood Ratio doubled in the role of the three-way interaction.

The *catmod* analysis for fatality crashes is shown in Table 11. The key term, the likelihood ratio, is marginally significant (*chi-square* = 3.02, *p*-value < 0.10). The

---

[8] This procedure is referred to as *backward-elimination*.

24

reduction in fatal crashes in rear impacts, relative to the reduction in fatalities in other impacts is quite high – the three-way Odds Ratio of 0.170 corresponds to a reduction of 83 percent (Table 12). The failure to yield a strongly significant result is due to the low number of fatalities in the dataset.

Table 11: CATMOD summary, fatality crashes

TERM	DF	CHI-SQ	p-value
Intercept	1	234.15	< 0.0001
Target	1	0.01	0.917
Urg	1	0.02	0.892
Likelihood Ratio	1	3.02	0.082

Table 12: Observed crash frequencies, fatality crashes versus all others

North Carolina	Rear Impacts	K	ABCN		
	1980 to 1993	3	74		
	1998 & newer	2	200		
				0.247	OR(1)
	Other Impacts				
	1980 to 1993	6	417		
	1998 & newer	26	1248		
				1.448	OR(2)
				0.170	OR(1) ÷ OR(2)

The *catmod* summary for the reduction in fatality and serious injury crashes is shown in Table 13. The odds ratio of 0.426 represents a reduction of 57 percent in fatalities and serious injuries. Again, the result is not statistically significant (Likelihood Ratio = 1.63, p-value > 0.10).

Table 13: CATMOD summary, fatality and serious injury crashes

TERM	DF	CHI-SQ	p-value
Intercept	1	322.85	< 0.0001
Target	1	0.33	0.569
Urg	1	4.16	0.042
Likelihood Ratio	1	1.63	0.202

Table 14: Observed crash frequencies, fatality and serious injury crashes versus all others

North Carolina	Rear Impacts	KA	BCN		
	1980 to 1993	6	71		
	1998 & newer	5	197		
				0.300	OR(1)
	Other Impacts				
	1980 to 1993	25	398		
	1998 & newer	54	1220		
				0.705	OR(2)
				0.426	OR(1) ÷ OR(2)

The above analysis is a replication as nearly as possible of the analysis from the Florida State Data System. Table 15 shows the results from Florida and North Carolina together. The reductions in K and KA for North Carolina may be spuriously high with such a small number of crashes.

Table 15: Summary of injury crash severity from Florida and North Carolina

Injury severity	Reduction in FL	Reduction in NC
K	27%	83%
KA	7%	57%

North Carolina – Supplementary data elements

Supplementary information was collected in the North Carolina special study beyond the State's standard police report – when the heavy vehicle was rear-impacted by a passenger vehicle, the following additional information was recorded:
- type of rear impact guard – no guard, old FHWA guard (60" wide), or new NHTSA / TTMA guard (94" to 102" wide)
- extent of passenger compartment intrusion in the passenger vehicle – none (rear impact guard successfully prevented an underride), underride without windshield damage, underride with windshield and/or A-pillar damage, underride with damage beyond windshield and into the roof
- orientation of striking vehicle – seven combinations of position (left, center, or right of the heavy vehicle) and angle (passenger vehicle approached from the left, from straight behind, or from the right)

Figure 3 (on page 3) is the portion of the special data collection form for recording the type of underride guard. Both graphic and textual descriptions were provided. The width of the underride guard is the primary identifying characteristic, and the presence of conspicuity tape is mentioned as a secondary characteristic that can distinguish between the two types of guards. The newer guards are also lower to the ground and set nearer to the rear of the trailer – these are important characteristics as well, but the forms were not designed to show all possible views.

First, the type of underride guard is presented, relative to the model years used previously. The accuracy is questionable because 45 of the MY 1998 and newer are said to have the older guard and 17 are said to have no guard or the old guard – 8 of these 17 are recorded as being either the van or flatbed type of trailer, which should be equipped with FMVSS compliant guards. These numbers may be a data collection inconsistency. But, if the MY 1998 & newer cohort includes a large proportion of special-use trailers that are exempt from the underride guard standard (e.g., wheels-back trailers), then this simple analysis by model year cohort could underestimate the effectiveness of the guards.

Table 16: Type of underride guard from special form based on model year of tractor

	No guard	Old guard	New guard
MY 1980 to 1993	10	44	25
MY 1994 to 1997	7	32	41
MY 1998 & newer	17	45	150

There are several reasons why these data are difficult to collect:
- The vehicle may have suffered damage to the underride guard during the crash.
- The vehicle may have suffered damage to the underride guard during some earlier incident (e.g., backing into a loading dock) and had not been repaired prior to the crash for which data is available.
- The ambient conditions were such that the highway patrol officer was unable to complete the form accurately (e.g., heavy traffic, poor lighting).

Figure 12 is the portion of the special data collection form designed to capture the extent of underride and the orientation of the striking angle of the passenger vehicle.

Figure 12: North Carolina special data collection form for extent of underride and striking orientation

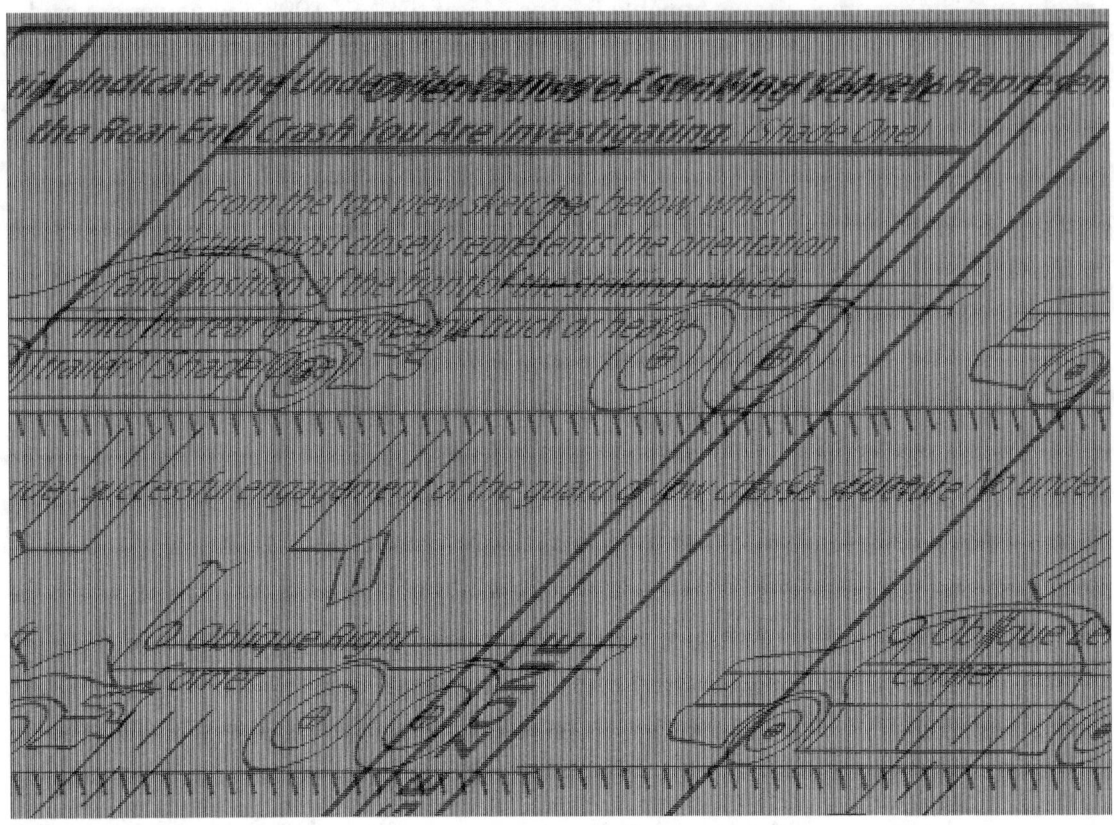

Table 17 shows the number of crashes for all combinations of damage zone and striking orientation. At right are the sums for the four corner orientations and the three center orientations. More than half of the corner crashes are "90 degrees left corner" (3), and nearly all of the center crashes are "90 degrees center" (7).

Table 17: Extent of underride and striking orientation, all possible combinations

Trailer MY 1980-1993	(1)	(2)	(3)	(4)	(5)	(6)	(7)	(1) to (4) Corner	(5) to (7) Center
No underride	1	0	5	4	0	1	31	10	32
Zone1 = minor	0	0	3	1	1	0	10	4	11
Zone2 = more severe	0	0	2	1	0	0	4	3	4
Zone3 = major	0	0	1	1	0	0	0	2	0
								19	47

Trailer MY 1994-1997	(1)	(2)	(3)	(4)	(5)	(6)	(7)	(1) to (4) Corner	(5) to (7) Center
No underride	1	1	12	5	1	3	23	19	27
Zone1 = minor	0	2	1	1	0	0	12	4	12
Zone2 = more severe	0	0	1	1	0	0	3	2	3
Zone3 = major	0	0	3	1	0	0	0	4	0
								29	42

Trailer MY 1998 & newer	(1)	(2)	(3)	(4)	(5)	(6)	(7)	(1) to (4) Corner	(5) to (7) Center
No underride	6	2	20	9	0	0	75	37	75
Zone1 = minor	2	1	10	3	2	1	30	16	33
Zone2 = more severe	0	1	7	0	0	1	5	8	6
Zone3 = major	1	1	3	0	0	0	1	5	1
								66	115

(1) Oblique left corner   (2) Oblique right corner
(3) 90 degrees left corner   (4) 90 degrees right corner
(5) Center oblique left   (6) Center oblique right
(7) 90 degrees center

Passenger compartment intrusion is more prevalent in corner impacts than in center impacts for any type of underride guard. For trailers with FMVSS-compliant guards that were impacted in the corner, there were 13 cases of severe or major intrusion, out of 66 total crashes where this information was recorded (13 of 66 = 19.7%). By comparison, there were seven center impacts with severe or major intrusion, out of 115 total crashes where this information was recorded (7 of 115 = 6.1%). The difference in the two proportions can be tested according to a binomial test. The statistical test is highly significant (p-value < 0.01), meaning that the result is not likely to be a chance occurrence owing to a small amount of data. It can be said that the center portion of the underride guard resists passenger compartment intrusion better than do the edge portions of the underride guard. FMVSS 223 requires a greater amount of force to be resisted near the center of the guard (locations P3 in Figure 2), compared to the edges (locations P1 in Figure 2).

Table 18 is a summary of the information contained in Table 17. For trailers with non-FMVSS-compliant guards (MY 1994-1997), the same relationship is found – namely, passenger compartment intrusion is more common in corner impacts compared to center impacts.

Table 18: Summary of passenger compartment intrusion by impact orientation and extent of intrusion

Trailer MY 1980-1993	(1) to (4)	(5) to (7)
	Corner	Center
Zones 2 & 3	5	4
Total	19	47
Percent 2 & 3	26.3%	8.5%

Trailer MY 1994-1997	(1) to (4)	(5) to (7)
	Corner	Center
Zones 2 & 3	6	3
Total	29	42
Percent 2 & 3	20.7%	7.1%

Trailer MY 1998 & newer	(1) to (4)	(5) to (7)
	Corner	Center
Zones 2 & 3	13	7
Total	66	115
Percent 2 & 3	19.7%	6.1%

For corner impacts, there is a small decrease in the percent of crashes with intrusion for Trailer MY 1980-1993 compared to FMVSS-compliant trailers (26.3% to 19.7%). For center impacts as well, there is a small decrease from 8.5 percent to 6.1 percent. These reductions are not statistically significant.

Of the total 11 fatalities and serious injuries in Table 14, seven crashes occurred when there was intrusion into Zone 2 or Zone 3, as shown in Figure 12. There are not enough crashes to draw statistical conclusions regarding the relationship between intrusion and injury outcome.

Consideration of intermediate model years 1994-1997

This report is primarily an analysis of the effectiveness of FMVSS-compliant underride guards (MY 1998 and newer) compared to those from earliest model years (1980-1993) that would most likely have been equipped only with the old-style FHWA guard that was narrower and set further from the rear of the trailer. It is not clear what effect the voluntary TTMA standard may have had because there was not an energy absorption and force resistance standard as in FMVSS 224.

Table 19 shows the observed crash frequencies from the three calendar year cohorts in Florida, along with those from the special study in North Carolina. The format is similar to that of Table 7. For all three calendar year cohorts in Florida, there is a relative increase in fatalities for the model years 1994 to 1997 compared to 1980 to 1993 – the three ratios are 1.545, 1.514, and 3.218. In North Carolina, the opposite is true – the ratio $OR(1) \div OR(2)$ of 0.328 represents a 67 percent reduction in fatalities.

Table 20 presents these same calculations for the reduction in fatalities and serious injuries (KA). Again, the three calendar year cohorts in Florida show a relative increase – the ratios of interest are 1.525, 1.099, and 1.404. In North Carolina, there is a reduction of 38 percent, based on the ratio presented of 0.615.

No conclusion can be drawn from these numbers because they differ in direction – relative increases in Florida, as opposed to relative reductions in North Carolina.

Table 19: Observed crash frequencies, fatality crashes versus all others, MY 1980 to 1993 versus MY 1994 to 1997

		Observed			
Florida					
CY 1993 to 2001	Rear Impacts	K	ABCN		
	1980 to 1993	16	423		
	1994 to 1997	5	174		
				0.760	OR(1)
	Other Impacts				
	1980 to 1993	35	869		
	1994 to 1997	8	404		
				0.492	OR(2)
				1.545	OR(1) ÷ OR(2)
CY 2002 to 2005	Rear Impacts	K	ABCN		
	1980 to 1993	15	436		
	1994 to 1997	13	432		
				0.875	OR(1)
	Other Impacts				
	1980 to 1993	48	2163		
	1994 to 1997	27	2106		
				0.578	OR(2)
				1.514	OR(1) ÷ OR(2)
CY 2006 & 2007	Rear Impacts	K	ABCN		
	1980 to 1993	2	106		
	1994 to 1997	5	111		
				2.387	OR(1)
	Other Impacts				
	1980 to 1993	11	368		
	1994 to 1997	11	496		
				0.742	OR(2)
				3.218	OR(1) ÷ OR(2)
North Carolina	Rear Impacts	K	ABCN		
	1980 to 1993	3	74		
	1994 to 1997	2	79		
				0.624	OR(1)
	Other Impacts				
	1980 to 1993	6	417		
	1994 to 1997	12	438		
				1.904	OR(2)
				0.328	OR(1) ÷ OR(2)

Table 20: Observed crash frequencies, fatality and serious injury crashes versus all others, MY 1980 to 1993 versus MY 1994 to 1997

		Observed				
Florida						
CY 1993 to 2001	Rear Impacts	KA	BCN			
	1980 to 1993	64	375			
	1994 to 1997	30	149			
				1.180	OR(1)	
	Other Impacts	KA	BCN			
	1980 to 1993	121	783			
	1994 to 1997	44	368			
				0.774	OR(2)	
				1.525	OR(1) ÷ OR(2)	
CY 2002 to 2005	Rear Impacts	KA	BCN			
	1980 to 1993	57	394			
	1994 to 1997	55	390			
				0.975	OR(1)	
	Other Impacts	KA	BCN			
	1980 to 1993	160	2051			
	1994 to 1997	138	1995			
				0.887	OR(2)	
				1.099	OR(1) ÷ OR(2)	
CY 2006 & 2007	Rear Impacts	KA	BCN			
	1980 to 1993	14	94			
	1994 to 1997	19	97			
				1.315	OR(1)	
	Other Impacts					
	1980 to 1993	42	337			
	1994 to 1997	53	454			
				0.937	OR(2)	
				1.404	OR(1) ÷ OR(2)	
North Carolina	Rear Impacts	KA	BCN			
	1980 to 1993	6	71			
	1994 to 1997	3	78			
				0.455	OR(1)	
	Other Impacts					
	1980 to 1993	25	398			
	1994 to 1997	20	430			
				0.740	OR(2)	
				0.615	OR(1) ÷ OR(2)	

DOT HS 811 375
October 2010

U.S. Department
of Transportation

**National Highway
Traffic Safety
Administration**